Cranberries

Nature's Nutrient Powerhouse

Candace Hoffmann

For permissions, ordering information, or bulk quantity discounts, contact: Woodland Publishing, 448 East 800 North, Orem, Utah 84097

Visit our Web site: www.woodlandpublishing.com
Toll-free number: (800) 777-2665

The information in this book is for educational purposes only and is not recommended as a means of diagnosing or treating an illness. All matters concerning physical and mental health should be supervised by a health practitioner knowledgeable in treating that particular condition. Neither the publisher nor the author directly or indirectly dispenses medical advice, nor do they prescribe any remedies or assume any responsibility for those who choose to treat themselves.

Cataloging-in-Publication data is available from the Library of Congress.

ISBN: 978-1-58054-461-0

Printed in the United States of America

Contents

As American as Cranberry Pie

The apple usually comes to mind as America's signature indigenous fruit, but did you know that the apple we know today was actually brought here from England? The American crab apple that greeted early settlers in the seventeenth century was dismissed as too bitter for British tastes, and seedlings of the more proper European apple were quickly requisitioned from England.

But while the newly transplanted pilgrims were lamenting the lack of a good apple, Native Americans were enjoying a wild fruit that *is* indigenous to North America—the cranberry.

Whether cranberries were part of the potluck dinner served up at what we now refer to as the first Thanksgiving is up for debate. Still, it is reasonable to suspect that not too long after the newcomers set foot in the New World they were introduced to the cranberry and the vibrant red fruit began showing up on dinner tables throughout the land.

The pilgrims dubbed the fruit "craneberry" because the plant's flower resembled the head, neck, and beak of a crane. According to the Cape Cod Cranberry Growers' Association, the settlers soon adopted craneberries for food and used them as coinage for bartering with their new neighbors. Native Americans held the cranberry in high regard both as a food source and for its medicinal uses.

In 1816, cranberries were successfully cultivated by Captain Henry Hall, and while there were other varieties of wild cranberries in Europe, the colonists soon began shipping their newfound product home.

Today cranberry production is limited mostly to the northern United States and Canada, with Wisconsin and Massachusetts yielding 56 percent and 26 percent, respectively, of the fruit found in grocery stores. Each year the United States produces some seventy thousand tons of cranberries, accounting for 82 percent of world production. Most of these berries are made into juice and sauces.

Some Cranberries Each Day May Keep the Doctor Away

But the cranberry is more than just a brightly colored holiday condiment. Recent and expanding research suggests that the cranberry may play a bigger role in keeping the doctor away than the apple.

Native Americans, who prized the cranberry for its beautiful red pigment and used it as a dye for rugs and blankets, were also aware of the berry's medicinal properties. According to Bradford Angier in his book *Field Guide to Medicinal Wild Plants*, Native Americans used the raw fruit as a poultice for dressing arrow wounds and brewed the plant's leaves into a tea as a tonic to help prevent kidney stones and "bladder gravel." Angier also notes that this cranberry tea was used as a diuretic, a treatment for kidney disorders, and as a cure for diarrhea, diabetes, and nausea.

The early New England settlers soon made good use of the healing powers of the little native fruit. According to the American Botanical Council, the use of cranberries to relieve blood disorders, stomach ailments, liver problems, vomiting, appetite loss, and cancer dates to the seventeenth century. The fruit became a staple on whaling and sailing ships when it was found to prevent scurvy—a good thing considering the lack of citrus trees in the American Northeast!

The Medicinal Value of Cranberries

Most people, however, are not aware of the full spectrum of health benefits that are derived from the cranberry. Research conducted by the University of Massachusetts, Dartmouth, randomly surveyed 260 consumers in the state and found that, of the 97 percent of respondents who said they've tried eating or drinking a cranberry product, 91 percent drink it because of taste, 68 percent because of health benefits, 38 percent because it is a holiday tradition, and 6 percent because it's available. Of those who reported that they were aware of the cranberry's health benefits, most respondents (41 percent—very aware; 46 percent—somewhat aware) knew that cranberry juice could help prevent urinary tract infections.

Beyond Bladder Health

There is good scientific evidence suggesting that the cranberry also:

- Promotes gastrointestinal health
- Promotes heart and artery health

- Has antiaging properties and may help prevent Alzheimer's dementia
- Helps in the recovery from stroke
- May help prevent cancer

Cranberry Nitty-Gritty

The botanical name of the cranberry grown for commercial use in the United States is *Vaccinium macrocarpon* Ait. of the Ericaceae (Heath) family. Close relatives include the mossberry, also known as the small cranberry (*Vaccinium oxycoccus* L.). Other relatives of the cranberry include the lingonberry (*Vaccinium vitis-idaea* L.); and within the genus *Vaccinium*, which includes some 150 to 450 diverse species, fall the blueberry, huckleberry, and the bilberry. The cranberry plant is also related to rhododendrons, azaleas, mountain laurel, heather, heath, and leatherleaf, which are part of the Ericaceae family as well.

Nature's Nutrient Powerhouse

The cranberry is a nutrient-rich superfood that has recently seen an uptick in scientific research scrutinizing its organic components—particularly its polyphenol content. Polyphenols, which include flavonoids, anthocyanins, and proanthocyanidins, have caught the attention of the research community because of their antioxidant, anti-inflammatory, and anti-adhesion properties. Moreover, according to research published in the January 2005 issue of the *American Journal of Clinical Nutrition*, "current evidence strongly supports a contribution of polyphenols to the prevention of cardiovascular diseases, cancer, and osteoporosis and suggests a role in the prevention of neurodegenerative diseases and diabetes mellitus."

Andrew Weil, MD, notes the increased interest in plant polyphenols, especially epigallocatechin gallate (EGCG), anthocyanins, and proanthocyanins. He explains that EGCG, in particular, shows efficacy in studies against cancer, is heart protective, and may also protect against UV radiation and precancerous sun damage.

What Exactly Are Antioxidants?

Antioxidants help rid the body of free radicals, which are byproducts of the oxygen we breathe. These destabilized oxygen molecules are missing an electron and their mission is to find and hook up with an oxygen electron. In their quest they often "steal" the needed electron from healthy cells, which causes an alteration in the "robbed" cells. Free radicals damage cellular DNA, proteins, and cell metabolism. In other words, our bodies are like a finely tuned machine, and free radicals are like sand and grit in the gears that make our machines malfunction.

Free radicals have been implicated in many diseases, including cancer, immune system disorders, elevated low-density lipoprotein (LDL) cholesterol levels, DNA damage, neurodegenerative disorders such as Alzheimer's disease, and even sunburn and wrinkles.

When discussing free radical formation, most experts suggest that the way apples turn brown when cut and exposed to air is a good example of oxidation at work. *Anti*oxidant action is evident when you dip a slice of apple in lemon juice. The apple stays white and fresh looking. Antioxidants work in your body in much the same way to keep free radical damage at bay.

Scoring High on the ORAC Scale

The quantity of antioxidants in a fruit or vegetable is most commonly measured according to their oxygen radical absorbance capacity (ORAC). Often you'll see a food's ORAC score or value referenced on food labels. Blueberries rank high with an ORAC score of 2,400. But cranberries are not far behind with an ORAC score of 1,750. Moreover, when total polyphenol content is considered, cranberries rank very high in comparison with other fruits, according to research cited by the Cranberry Institute.

Polyphenol Content by Standard Serving

• Pure cranberry juice	587 mg
• Red wine	400 mg
• Grape juice	356 mg
• Apple juice	61 mg
• Cranberry juice cocktail (27 percent juice)	137 mg
• Cranberries (55 grams)	373 mg
• Red grapes (140 grams)	518 mg

High–ORAC value foods, researchers are learning, have many health benefits, including helping to slow the aging process. In essence, they keep your body humming along and in tune.

Tufts University researchers have found evidence that foods with a high ORAC value increase the antioxidant power of the blood and suggest that choosing fruits and vegetables that will give you an intake of between 3,000 and 5,000 ORAC units daily can significantly impact plasma and tissue antioxidation capacity. With their high ORAC value, cranberries can play an important part in your daily tally of ORAC units.

Bioactive Constituents in Cranberries

Cranberry's polyphenols, as well as its vitamin and fiber content, all add up to one nutrient-rich red berry! Many of the following chemical components are polyphenols—phytochemicals that are eliciting a lot of attention from researchers who are studying and proving the many medical benefits of the cranberry.

- Anthocyanidins
- Benzoic acid
- Chlorogenic acid
- Ellagic acid
- Epigallocatechin gallate (EGCG)
- Fiber
- Flavonoids
- Flavan-3-ols (catechins)
- Folate
- Lignans
- Myricetin
- Potassium
- Proanthocyanidins
- Quercetin
- Resveratrol
- Thiamin
- Triterpenoids
- Vitamin A
- Vitamin C

Breakthroughs in Cranberry Research

Amy Howell, PhD, has been working in cranberry research for some fourteen years, and may well be a cranberry guru. She is an associate research scientist at the Philip E. Marucci Blueberry and Cranberry Research and Extension Center at Rutgers University. "Can women be gurus?" she asked during an interview. If they can, Howell is certainly a guru in the world of cranberries.

She said she was fascinated by the berry because "it is one of the few fruits with a lot of anecdotal evidence. There was a big body of folklore, people knew it worked and have been drinking cranberry juice for hundreds of years."

Early research to understand the mechanism behind the cranberry's many health benefits, particularly its ability to prevent urinary tract infections, led to a hypothesis by German physicians in the nineteenth century that perhaps it was the cranberry's hippuric acid content that gave it its medicinal punch. Hippuric acid is known for its antibacterial capability, and this hypothesis formed the basis of the long-held conclusion that cranberry ingestion made urine more acidic and thus helped ward off infection.

Howell noted that the first journal article on cranberries and urinary tract infections was published in 1906, bolstering the idea that it was the fruit's acidity that helped prevent infection. "But if that is the reason, why doesn't grapefruit, lime, or orange juice help with UTIs?" Howell asked.

Clinical studies in the late 1980s and 1990s showed that it wasn't the acidity of cranberries that kept UTIs at bay, but something else, Howell said. Researchers also discovered that cranberries had an anti-adhesion effect. In other words, the bacteria that cause UTIs didn't stick to the bladder lining and were simply washed away.

But still, the mechanism of action in this process had not yet been identified. Howell took up the gauntlet, and after five years of research found what may be the answer—proanthocyanidins (PACs). These compounds, Howell said, are responsible for the cranberry's efficacy in preventing UTIs.

In vitro (in test tube or petri dish) studies suggest that cranberry proanthocyanidins prevent the adhesion of *Escherichia coli* (*E. coli*) to the cells lining the bladder (uroepithelial cells). *E. coli* is one of the most common bacteria responsible for causing UTIs. In short, before an infection can take hold, cranberry juice flushes the bacteria away.

Cranberry's Unique Proanthocyanidins

Howell explained that proanthocyanidins, which are condensed, or high-molecular-weight tannins, are high in antioxidant activity. Tannins are also found in tea, chocolate, and red wine. However, cranberry's particular tannins stood out from the crowd upon closer scrutiny. "When we isolated them from the cranberries and looked at the shapes of the molecules compared to other foods, we found that they are really different. They have an extra link or bond—two bonds instead of one," Howell said. "This is very rare. We could not find this in chocolate, grape, or tea tannins." Howell and her research team soon discovered that very few tannin molecules have these bonds, which are called *type A linkages*, and they concluded that cranberry's unique PACs are responsible for this anti-adhesion effect.

Get Your PACs Here

Cranberry's unique anti-adhesion and antibiotic proanthocyanidins are distinctive among fruits. Whether you eat cranberries fresh, dried, in a sauce, or in cranberry juice cocktail, you can incorporate these valuable antioxidants into your diet. The following quantities all deliver similar quantities of PACs:

- 10 ounces of cranberry juice cocktail
- 1 1/2 cups of fresh cranberries
- 1 once of dried and sweetened cranberries
- 1/2 cup of cranberry sauce

SOURCE: The Cranberry Institute

Prevention of Urinary Tract Infections

The use of cranberries to treat UTIs, which began as a folk remedy, is standing up well under intense scientific scrutiny by Howell and other researchers. Consider these statistics from the American Urological Association: Urinary tract infections are responsible for over seven million doctors' office visits each year and about 5 percent of all visits to primary care physicians. More women (40 percent) than men (about 12 percent) will experience at least one symptomatic UTI in their lifetime. Women will have 25 percent more urinary tract infections than men, and between the ages of twenty and fifty will have fifty times more UTIs than men, according to *The PDR Family Guide to Women's Health and Prescription Drugs*.

Preventing UTIs is paramount, especially for those who are prone to them, and regular consumption of cranberry appears to help. In fact, clinical trials have been fairly consistent in showing that about 50 percent of people with UTIs are helped by drinking cranberry juice, Howell is quoted as saying in the June 2005 issue of the *Nutrition Action Health Letter*.

The National Kidney and Urologic Diseases Information Clearinghouse (NKUDIC) offers these additional measure that women can take to prevent UTIs:

- Drink plenty of water every day
- Urinate when you feel the need; don't resist the urge to urinate
- Wipe from front to back to prevent bacteria around the anus from entering the vagina or urethra

- Take showers instead of tub baths
- Cleanse the genital area before sexual intercourse
- Avoid using feminine hygiene sprays and scented douches, which may irritate the urethra.

UTIs in men, the experts at the NKUDIC note, are often caused by a blockage of some sort, such as a urinary stone or an enlarged prostate (benign prostatic hyperplasia). In older men, UTIs are often associated with an acute bacterial infection of the prostate—bacterial prostatitis. These are urgent problems and must be treated by a physician as they can have adverse and serious consequences.

Cranberry Juice, Immunity, and UTIs

Once a urinary tract infection takes hold, medical attention is required and a course of antibiotics is the preferred treatment. So far, the medical literature does not show that cranberries are an effective treatment for UTIs, but some experts suggest that continuing to drink cranberry juice as an adjunct to antibiotic treatment can do no harm and may help bolster your immune system.

Howell also emphasized the importance of incorporating cranberries into your daily diet, especially if you are prone to recurrent UTIs. She said that "your best bet is to drink two glasses of a cranberry beverage, one in the morning and one before you go to bed," because the effects wear off after about ten hours.

Antibiotic Resistance

Another good reason to prevent UTIs with cranberry juice is antibiotic resistance. Bacteria are living organisms that go through an evolutionary process just as we do. When they encounter an antibiotic, some of the stronger bacteria do not die. They multiply and their offspring become "resistant" strains, meaning they are no longer killed by that particular antibiotic. This is why it's unwise to discontinue a course of antibiotics, even if you're feeling better, before you've taken all of the prescribed medication. One should also avoid taking antibiotics at the slightest sniffle or for viral infections, which antibiotics cannot treat.

The North American Urinary Tract Infection Collaborative Alliance (NAUTICA) studied the problem of antibiotic resistance in United States and Canadian outpatients by examining urine samples. They found that the most common bacteria were *E. coli* (57.5 percent) and *Klebsiella pneumoniae* (12.4 percent). They tested commonly used antibiotics against the bacteria and found that 45.9 percent of the bacteria were resistant to ampicillin (Omnipen, Polycillin, Principen) and 20.4 percent were resistant to SMX/TMP (Septra, Bactrim, Sulfatrim), one of most common and most effective UTI treatments. Resistance to other commonly used antibiotics—nitrofurantoin (Furadantin, Macrobid, Macrodantin), ciprofloxacin (Cipro, Cipro XR), and levofloxacin (Levaquin)—was also found, albeit to a lesser extent.

E. coli *Makes Society Hit List*

More recently, the Infectious Diseases Society of America released its "Hit List of Dangerous Bugs," and *E. coli* and *Klebsiella* were among the six listed bacteria that have become resistant to current antibiotic treatment. In a March 1, 2006, press release the IDSA summarized:

> These bacteria are major causes of urinary tract, gastrointestinal tract, and wound infections. They are becoming resistant to a growing number of antibiotic classes at the same time as the frequency of outbreaks is increasing. Failure to treat with the appropriate antibiotics during a recently documented *K. pneumoniae* outbreak increased the mortality rate from 14 percent to 64 percent. . . . New therapies are badly needed.

While not a substitute for antibiotics, cranberries could go a long way in keeping germs away! A study from the University of California, Irvine, reported in the November–December 2005 issue of *Renal and Urology News*, found that 50 to 64 percent of people who regularly consumed cranberry capsules showed antimicrobial activity against *E. coli, Klebsiella pneumoniae,* and *Candida albicans* (a yeast that causes vaginal infections).

And, consider this research from scientists in Turkey: Writing in the journal *Mediators of Inflammation,* Ergul Belge Kurutas and colleagues found that there was a lowering of antioxidant activity in the presence of UTIs. They write in their conclusion: "We believe that patients with UTI may benefit from antioxidant treatments in addition to antibacterial treatment."

It's better to avoid an antibiotic-resistant infection in the first place than it is to treat the infection later! Just two glasses of cranberry juice a day can keep you on the road to urinary tract health.

Efficacy of Cranberry Capsules and Dried Cranberries

If you can't drink cranberry juice every day—or if you just don't like the taste or don't want the additional calories—cranberry nutritional supplements are widely available. Dr. Howell said that the supplements are effective, and she recommends reading the labels to ensure that any supplement you're considering is made from whole fruit.

Dried cranberries also have been shown to protect against the adhesion of *E. coli*, Dr. Howell said. A small pilot study shows that dried cranberries do demonstrate anti-adhesion activity against *E. coli* and suggests that the dried version of the fruit may be a viable alternative to cranberry juice for the prevention of UTIs. However, until more evidence is gathered, dried cranberries might be best used as an additional way of getting your invaluable PACs along with cranberry juice, especially if you are prone to UTIs.

Beyond UTIs

The anti-adhesion effect of cranberries is doing far more than protecting your bladder. It stands to reason that if they help prevent the adhesion of bacteria to the lining of the bladder, this effect may work in other areas of the body. In fact, a study from the Center for Bio/Molecular Science and Engineering, Naval Research Laboratory, corroborated Howell's findings about cranberry's anti-adhesion capabilities in a study published in the February 2006 issue of *Annals of Chemistry*. The authors write: "This [anti-adhesion capability] cannot be duplicated with grape, orange, apple, or white cranberry juice."

Cranberries and Gastrointestinal Health

With a better understanding of cranberry's mechanism of action in preventing *E. coli* bacteria from adhering to the urethral endothelia, researchers are looking into the juice's effect on preventing the adhesion of another bacterium, *Helicobacter pylori (H. pylori)*, which, the American Gastroenterological Association (AGA) says, is implicated in many gastrointestinal disorders. *H. pylori* is nearly always present in stomach inflammation, in the vast majority of duodenal and gastric ulcer patients, in about 50 percent of people with heartburn, and in gastric cancer patients.

The shape of the bacterium facilitates its movement though the mucus layer of the stomach and duodenum. *H. pylori* produces the enzyme urease, which the AGA notes, "helps it survive in the hostile acidic environment of the stomach and disrupts the mucus layer structure. [The bacteria] produces chronic infection, and once a person is infected—usually in childhood—it's probably for life." While prevalent in developing countries, *H. pylori* is found worldwide, although there does appear to be an inverse relationship between infection and economic status, the AGA says.

While there is a 20 to 50 percent incidence of *H. pylori* infection among adults worldwide, there is more than an 80 percent incidence in developing countries, scientists at Beijing's Institute for Cancer Research observe in their research paper on using cranberry juice as a prophylactic (preventive agent) against *H. pylori*. Since the bacteria seem to be associated with an increased risk of developing gastric cancer, preventing *H. pylori* infection is a standard of care, usually with a multiple antibiotic regimen. However, as stated before, antibiotic resistance and the side effects associated with their use, as well as the huge costs associated with such a treatment, make prophylactic antibiotic use against *H. pylori* undesirable, the researchers note. Cranberry juice, however, presents a cost-effective and safe strategy in preventing *H. pylori* infection, the Chinese scientists found.

L. Zhang and his colleagues randomized 189 participants to receive two 250 milliliter (eight ounce) doses of either cranberry juice cocktail or a placebo daily for ninety days. Using breath testing to analyze the *H. pylori* infection status of the subject, they found that participants who ingested the cranberry juice had a 14 percent reduction in the rate of *H. pylori* infection.

The participants were from Lunqu County in rural China, which has the world's highest incidence of stomach cancer. *H. pylori* is seen in 52 percent of the children and 72 percent of adults. "[W]e observed that cranberry juice can retard *H. pylori* infection in humans and may be a promising new form of therapy for worldwide management of this infection that does not induce the side effects caused by antibiotics," the study authors write.

This is the first study to show that regular consumption of cranberry juice can have this effect in preventing *H. pylori* infection.

Another study, reported in the journal *Applied and Environmental Microbiology,* looked at the effect of a combination of extracts from cranberries and oregano on *H. pylori* and found that the combination showed more efficacy against the bacteria than extracts of either substance alone, suggesting synergy between the two. "These combinations of beneficial plant extracts provide a natural and dietary solution, as well as an additional strategy to inhibit the growth of *H. pylori*," the researchers from the University of Massachusetts, Amherst, wrote in their report.

As the Chinese study showed, just two juice box–size servings of cranberry juice cocktail can go a long way in preventing *H. pylori* from taking hold. But what if you're already suffering with an *H. pylori*–mitigated ulcer? Cranberry juice may still help. A study conducted at the Helicobacter Pylori Research Institute in Petah Tikva, Israel, suggests that combining antibiotics and a cranberry preparation may help combat *H. pylori*.

Ulcers, like UTIs, can be prevented if the bacteria that cause them cannot take hold. Daily consumption of cranberry juice is as good for your gastrointestinal health as it is for your bladder health.

Note: Ulcers can also develop if you habitually use pain relievers known as nonsteroidal anti-inflammatory drugs (NSAIDs) such as aspirin or ibuprofen (Motrin, Advil).

Cranberries—The New Cavity Fighter?

In his book *Folk Remedies from Around the World*, John Heinerman offers this solution to bleeding gums: "If you have access to some of these wild or cultivated cranberries in their fresh state, crush a few of them and then rub the mixture across the gums with your forefinger after every brushing. This will halt the bleeding. Or else thaw a few frozen cranberries, mash them well with the back of a spoon, and do the same thing." He also suggests taking the contents of a commercially available cranberry capsule and rubbing that across your gums to stop the bleeding.

Just folklore? Possibly not.

Consider the research being conducted by biologist Hyun (Michel) Koo, DDS, PhD. He believes that cranberry juice could do for teeth what it's doing for UTIs. In a press release from the University of Rochester Medical Center, where he conducted his research, he noted: "Scientists believe that one of the main ways that cranberries prevent urinary tract infections is by inhibiting the adherence of pathogens on the surface of the bladder. Perhaps the same is true in the mouth, where bacteria use adhesion molecules to hold onto teeth."

The case in point this time is the bacterium *Streptococcus mutans*, which "is generally regarded as a primary microbial culprit in the etiology of dental caries because of its acidogenic and aciduric [associated with acid tolerance] properties together with its ability to synthesize extra cellular glucans from sucrose by glucosyltransferases (GTFs)," Koo and his colleagues write. It's not just sugar, but the way sugar and bacteria team up that allows cavities to form.

Koo and his colleagues found that a beverage containing 25 percent cranberry juice can block the bacteria that form plaque, and he sees definite possibilities for its use. "Cranberry is a promising natural product with multiple inhibitory effects against several virulence factors associated with the pathogenesis of dental caries," he writes in his conclusion.

Researchers in Israel are also studying cranberry components and are finding them effective against oral bacteria, but note that commercial juice is too high in dextrose and fructose for oral use and that more study is needed in this area. But don't be surprised if a cranberry-based oral rinse finds its way onto supermarket shelves in the not-too-distant future. In the meantime, Heinerman's folk remedy might be worth a shot.

Cranberries and Heart Health

Cardiovascular disease, the number-one killer of American men and women, claims more victims each year than cancer, chronic lower respiratory disease, accidents, and diabetes combined, the American Heart Associations says.

The best news about the bad news of heart disease is that it is largely preventable through lifestyle changes—namely diet and exercise. One cannot discount the importance of exercise in keeping your body's most important muscle—your heart—in tiptop shape. However, the way we eat, and eat, and eat unfortunately takes its toll. Not only does our modern convenience-food lifestyle include too much fat, sodium, and sugar, it is often nutrient deficient as well, lacking the essential vitamins and minerals we need to maintain our health.

The major modifiable risk factors for heart disease, the American Heart Association says, include tobacco use, high blood cholesterol, high blood pressure, physical inactivity, obesity, and diabetes.

Atherosclerosis, or hardening of the arteries, results from the buildup of fatty deposits (LDL, or "bad" cholesterol) that harden into a plaque on artery walls. It is a leading cause of cardiovascular disease, often causing heart attack and stroke. The plaque, the result of LDL oxidation, causes narrowing of the arteries that restricts blood flow. Additionally, pieces of the hardened plaque can break off and lodge in another part of an artery, completely blocking blood flow, resulting in a heart attack, aneurysm, or stroke.

The abundance of polyphenols in cranberries, particularly flavonols, anthocyanins, and proanthocyanidins, could play a part in preventing heart disease. Researchers from the University of Wisconsin–Madison have found that these polyphenols inhibit the oxidation of LDL cholesterol.

Researchers at Cornell University in Ithaca, New York, who evaluated cranberries' antioxidant capacity found in their in vitro experiments that cranberries exhibited potent activity against LDL oxidation. One hundred grams of cranberries produced an antioxidant capacity equivalent to 1,000 mg of vitamin C, or 3,700 mg of vitamin E. Cranberries also "significantly induced expression of hepatic (liver) LDL receptors and increased intracellular uptake of cholesterol in HepG2 cells in vitro in a dose dependent manner," the researchers write. LDL cholesterol is cleared from the body through the liver, and this is how statin drugs work.

To further strengthen the case for making cranberries a regular part of our diets, Joe Vinson, PhD, a researcher at the University of Scranton, has been testing the antioxidant value of cranberry juice and measuring its effect on study participants' lipid profiles. While he did not find that cranberry juice actually lowers LDL cholesterol, he found something that may be equally important: it raised HDL cholesterol, the "good" cholesterol that helps break down and move LDL cholesterol out of our bodies.

Participants in Vinson's study consumed three servings (240 milliliters, or about eight ounces) of cranberry juice a day for one month. Their HDL cholesterol increased 10 percent, which medical experts say decreases heart disease risk by 40 percent! Vinson also found the following: Cranberry juice consumption at high doses, even with sugar, does not lead to weight gain.

Long-term consumption of cranberry juice provides an increase in plasma antioxidant capacity of up to 121 percent and provides a 36 percent reduction in lipid peroxidation products.

In the study, however, triglycerides were elevated with medium and high doses, although this could be related to the high-fructose corn syrup in the sweetened cranberry juice cocktail. The artificially sweetened cranberry juice did not show this effect.

"One or two glasses a day of cranberry juice are about what is needed for UTI prevention, and it's probably also good for the heart," Dr. Vinson said. "Our research parallels that."

Another study presented at the thirty-fifth congress of the International Union of Physiological Sciences in April 2005 showed that regular consumption of cranberry juice over six months improved vascular function in animals (pigs) with high cholesterol and atherosclerosis. "Since the abnormal functioning of blood vessels is an important component of heart disease, finding a way to improve vascular function in patients with high cholesterol and atherosclerosis is critical to helping protect these patients from consequences such as heart attack or stroke," lead researcher Kris Kruse-Elliot said in a prepared statement on the study. While this was an animal study and the human equivalent intake of cranberry juice would be either four to eight servings of dried cranberries or ten to twenty servings of cranberry juice, research collaborator Jess Reed notes in the release: "[T]he point to be emphasized is that total polyphenol intake is very low in western diets and a diet rich in polyphenols would in fact give a daily intake that is equivalent to the levels in our cranberry feeding experiments."

Killing Two Diseases with One Berry—The UTI Connection

The American College of Cardiology defines acute coronary syndrome (ACS) as a collective term for any of the signs and symptoms suggestive of acute myocardial infarction (heart attack) or unstable angina (chest, arm, or jaw pain). Is there a connection between ACS and UTIs?

Researchers from the University of Texas Southwestern Medical Center think there may be. Inflammation is implicated in the development of atherosclerotic disease, and the researchers point out that conditions that cause systemic inflammation, such as a subclinical infection, could trigger ACS. They evaluated the prevalence of UTIs among one hundred ACS patients and compared them with a group of the same age who were undergoing coronary artery bypass graft surgery. They found that UTIs were common among patients with ACS and write: "This hypothesis should be explored in other data sets, and similar relationships with other bacterial and viral infections should be examined."

If there is a true UTI connection, preventing UTIs, and possibly even ulcers, with regular consumption of cranberry juice may in the long run also protect your heart.

Inflammation

Inflammation is implicated in a number of disease processes including atherosclerosis and cancer. Cranberry's polyphenols have proven anti-inflammatory properties as well as antioxidant and anti-adhesion profiles. A study from the University of Illinois at Chicago found that the resveratrol in cranberries ranks close to that of wine and grape juice. Controlling inflammation is important in preventing and treating heart disease, cancer, arthritis, autoimmune disorders, and many other diseases and conditions.

Cranberries and Cancer

Cancer is the monster disease everyone dreads, but few understand. Cancer occurs when the body's normal cycle of new cell growth and old cell death, which helps constantly renew and regenerate our tissues, goes awry. Apoptosis, or programmed cell death, is vitally important. Without this programming, our cells, even normal ones, would grow out of control.

In the past, it was often thought that cancer happens mostly to those who are genetically predisposed to it, but this turns out to be incorrect. Medical science, the American Cancer Society says, now understands that environmental factors—tobacco use, poor nutrition, inactivity, obesity, certain infectious agents, certain medical treatments, sunlight, and carcinogens in food, air, water, and soil—all put us at risk for cancer. Thankfully, the most common causes of cancer—tobacco use, poor nutrition, inactivity, obesity, and sunlight—are modifiable, and adopting a healthy lifestyle can help keep us cancer-free.

This is where consumption of nutrient-rich cranberries comes in. Scientists studying cranberry compounds found that the proanthocyanidins, the same polyphenols responsible for preventing the adhesion of *E. coli* to the bladder lining in the prevention of UTIs, also inhibits the growth of colon and leukemia cancer cells in vitro. "While previous studies have shown that cranberry extracts inhibit the proliferation of cancer cells, this is the first study to confirm that it's the cranberry PACs that are the active components," Catherine C. Neto, lead author of the study and associate professor in the department of chemistry and biochemistry at the University of Massachusetts, Dartmouth. "This study is a significant step toward helping to establish a body of research that shows cranberry PACs may also work to prevent tumor cell growth in vivo."

Researchers at the David Geffen School of Medicine at UCLA wanted to see exactly what accounts for cranberries' effect against cancer, so they analyzed and separated total cranberry extract (TCE) into its component fractions, citing other studies that have attributed cranberry's anti-adhesion action to different groups of compounds—organic acids, sugars (e.g., fructose) and proanthocyanidins. They also compared the anti-cancer-proliferation activities of these individual fractions against TCE to see if there were any synergistic effects among various component combinations. "This is the first report of the human tumor cell inhibitory activities of cranberries against this panel of oral, prostate, and colon cancer cell lines in a highly sensitive luminescent ATP [sensitive way of testing antiproliferative activity] cell viability assay," the study's authors write.

The researchers found activity against all the cancer cell lines and that the TCE was more effective than the components alone. They also noted that the cranberry sugars did not inhibit the proliferation of any of the cancer cell lines. The study authors conclude:

It is well established that consumption of fruits and vegetables has been associated with reduced risk of chronic diseases such as cancer and that plant extracts including fruits and berries show antitumor activities. Our studies have shown that enhanced antiproliferative activity is obtained when cranberry extract is enriched in polyphenolic content by removing fruit sugars and organic and phenolic acids. Also, there were additive or synergistic antiproliferative effects resulting from the combination of anthocyanins, proanthocyanidins and flavonol glycosides compared to individual purified phytochemicals. The observed antiproliferative activities of cranberry phytochemicals against tumor cells provide some basic evidence for the potential anti-cancer effects of cranberry polyphenols and suggest that studies of cranberry extracts should be carried out in appropriate animal models of cancer and ultimately in human cancer prevention trials.

Researchers at the University of Western Ontario suggest that cranberry extracts could yield a new cancer drug after they found that an extract of cranberry press cake demonstrated efficacy against several tumor cell lines, including breast, prostate, skin, colon, lung, and brain, when injected in mice.

While it may be a while before a cranberry-derived cancer treatment is developed, whole cranberries, whether fresh, dried, juiced, or sauced, are a good addition to an optimally nutritious diet, which is one of the mainstays of cancer prevention.

The American Cancer Society recommends the following lifestyle changes to lower your risk for cancer:

- Eat a variety of healthful foods with an emphasis on plant sources
- Eat five or more servings of a variety of fruits and vegetables each day
- Use whole instead of processed/refined grains and sugar
- Limit consumption of red meats, especially high-fat and processed meats
- Choose foods that help maintain a healthy weight
- Exercise regularly—at least thirty minutes of moderate activity five or more days a week
- Maintain a healthy weight throughout life by balancing caloric intake with physical activity; lose weight if you are currently overweight or obese
- Limit alcoholic drinks to no more than two drinks per day for men; one drink per day for women

Cranberries and Neurological Disorders

Research into Alzheimer's disease and other neurodegenerative diseases, such as amyotrophic lateral sclerosis (ALS; Lou Gehrig's Disease) and Parkinson's disease, is also pointing a finger at the possible role that oxidative stress—free radical damage—plays in development of changes in the brain.

To help ward off neurodegenerative disorders, experts recommend maintaining optimal health with the addition of fruits and vegetables, which are believed to prevent aging of both the body and the brain. At the Jean Mayer USDA Human Nutrition Research Center on Aging at Tufts University, scientists cite oxidative stress as one of the major abuses the brain endures as we age, because it results in the "inability to balance and to defend against the cellular generation of reactive oxygen species (ROS). These ROS cause oxidative damage to nucleic acid, carbohydrate, protein and lipids," the scientists write in the journal *Neurobiology and Aging*. Such damage is particularly detrimental to brain cells resulting in many of the normal declines seen with aging in cognition and motor performance.

Diseases such as amyotrophic lateral sclerosis (ALS), Alzheimer's disease, and Parkinson's disease exacerbated these problems. The researchers found that polyphenols found in berries, such as blueberries, can reverse age-related neuronal decline.

It isn't a far stretch, given cranberries' high polyphenol content, to conclude that eating and drinking cranberry products may help in this context as well.

Another of cranberry's polyphenols is epigallocatechin gallate, which is gaining scientific acceptance for helping to prevent neurological disorders. EGCG from green tea was shown in a Korean study to have potential therapeutic uses as a treatment for ALS.

Resveratrol, another component of cranberries, was found in a study that examined the compound in wine, to have neuroprotective properties since regular consumption of wine is associated with a lower incidence of Alzheimer's disease, and researchers in Korea suggest that resveratrol may also have therapeutic potential in this context.

One pilot study from the Virginia Polytechnic Institute and State University in Blacksburg, Virginia, tested the short-term effect of cranberry juice on the neuropsychological functioning of cognitively intact older adults. The double-blind, randomized, placebo-controlled study did not show any significant relationship between cranberry juice consumption and cognitive processes, moods, energy levels, and overall health. The study did, however, produce one interesting finding. Twice as many of the study participants who received the cranberry juice rated their overall memory abilities at the end of the treatment as "improved" compared with those in the control group who did not receive the juice.

The study's authors, headed by W. David Crews Jr., PhD, of Virginia Neuropsychology Associates, say in their conclusion that more study over longer periods of time, with a larger consumption of cranberry juice versus placebo, is needed "in an effort to more closely parallel the previous research involving *Vaccinium* products and aged, laboratory animals." Science is finding that oxidative stress is so closely aligned with neurological disorders that Crews and his colleagues are probably intuitively on the mark.

Researchers from the University of Virginia Health System are finding that in Parkinson's disease the oxygen free radicals "are damaging proteins in mitochondria, the tiny cellular 'batteries' of brain cells," and conclude it may be one of the main causes of the disease. The authors hypothesize that drugs that could limit damage from free radicals could help. Perhaps a diet rich in antioxidant-containing foods, such as cranberries, while certainly not a cure for this devastating disease, might at least help prevent some of the oxidative damage.

Oxidative stress shows up in a lot of papers discussing neurobiological diseases and conditions. While fledgling in our understanding of exactly what is going on in the brain both in the disease process and the role of free radicals in it, antioxidants are piquing keen interest in the scientific community.

Cranberries and Rheumatoid Arthritis

Rheumatoid arthritis, a severe, debilitating autoimmune disease, may also have a relationship with urinary tract infections caused by another bacterium: *Proteus mirabilis.* A study conducted at King's College London suggests that people in the early stages of rheumatoid arthritis might benefit from "*Proteus* antibacterial measures involving the use of antibiotics, vegetarian diets and high intake of water and fruit juices such as cranberry juice in addition to the currently employed treatments."

Cranberries, the Common Cold, and Influenza

Instead of reaching for that glass of orange juice the next time you get a cold or flu, you might consider cranberry juice. Its now well-documented antibacterial activity makes it a perfect choice for bolstering your immune system when you're confronted with the onslaught of cold and flu viruses. And, with the on-again, off-again availability of influenza vaccine, adding the most nutritious foods and juices to your diet can be the front line of your defense.

Researchers in Israel have found that cranberry juice constituents appear to have an inhibitory effect on influenza virus adhesion and infectivity, and suggest it may have therapeutic potential. While their research is still in the laboratory, watch this area for more investigations. Or, experiment on yourself and increase your intake of cranberry juice the next time you catch a cold or the flu. Unlike many drugs, there are few potential side effects associated with drinking cranberry juice.

Possible Adverse Side Effects

While cranberries are relatively safe, taking too much of anything, even healthful things, has the potential to cause problems.

Large amounts of cranberry, beyond normal food serving sizes, could result in diarrhea and other gastrointestinal symptoms, according to the National Center for Complementary and Alternative Medicine. The center also notes that the safety of cranberries in amounts greater than consumed in foods is unknown.

Interaction with Warfarin

Outside of not going on a cranberry-only diet or taking cranberry supplements beyond the recommended dosages, a possible problem is the interaction of cranberries with the anticoagulant drug warfarin. People taking warfarin (Coumadin) should not take cranberry in any form without checking with their doctors. Also, for people who have a history of calcium oxalate kidney stones, some research suggests that cranberry consumption may promote the formation of this type of stone. Although, as noted in *Better Nutrition*, a 2003 study indicated that cranberry actually reduced the incidence of such stones. But since kidney stones create such an excruciatingly painful experience, it might be best to err on the side of caution.

Adding Cranberries to Your Diet

While one would optimally want to drink pure cranberry juice, the reality is that it would "rip the enamel off your teeth," Dr. Howell said. "The tannin [which supplies its bitter, sour flavor] is produced as a plant defense to keep birds and predators from eating the fruit."

Howell suggests mixing pure cranberry juice with other juices, such as grape juice (which will give a bigger ORAC bang for your buck) or adding the pure juice as a splash to mixers.

The good thing about cranberries is that in all forms their PACs are not affected, Dr. Howell explains. Plus, fresh cranberries are easily stored. Experts at the University of Wisconsin–Madison say that cranberries may be frozen in their plastic packaging for up to nine months, and some sources say for as long as several years. It is best to keep cranberries frozen until you want to use them because they become mushy when thawed and are more easily chopped or ground while frozen. While many sources will recommend specific amounts of cranberries for specific ailments, the main thing is to begin incorporating cranberry juice or the whole fruit into your diet to reap its antioxidant and antibacterial benefits.

Culling information from studies and various sources, we can make several recommendations:

- People with recurrent UTIs may want to drink two to four eight-ounce glasses of cranberry juice (or cocktail, artificially or sugar-sweetened), half in the morning and half in the evening.
- To prevent gastric ulcers caused by *H. Pylori*–type bacteria, try two eight-ounce glasses of juice per day.
- For normal prevention and overall health, drink two to four eight-ounce glasses of juice per day.

Cranberry Supplements

A very convenient way to ensure that you get your daily cranberry boost is to take cranberry nutritional supplements, either as capsules or tinctures. The recommendation is to take supplements made from whole cranberry juice extract and take one 400-milligram capsule a day. In tincture form (1:5 concentration), the dosage would be a half-teaspoon three times daily.

Dried and Fresh

Don't forget that dried and fresh cranberries can help supplement your juice intake. Use dried cranberries as you would raisins—in your cereal, in your oatmeal, as a nice addition to salads, puddings, or yogurt.

Fresh cranberries make delicious relishes and can add a nice tart flavor to salads and other recipes where a blend of tart and sweet is desired, such as in fruit salsas.

Conclusion: Cranberries and Optimal Health

Cranberries are such a nutrient-rich food that they shouldn't be relegated to holiday dinners. Find ways to use them all year long. Science and, more importantly, research funding, is just catching up to what Native Americans knew all along—cranberries pack a walloping health benefit.

References

Ackerson, Amber D. "Cranberry (*Vaccinium macrocarpon*)" *Better Nutrition* 67; no. 5 (May 2005) 12.

American Gastroenterological Association. "Peptic Ulcer Disease" and "Peptic Ulcer Disease and *H. Pylori*." www.gastro.org. Accessed 2/22/2006.

American Physiological Society. "Cranberry juice modulates atherosclerotic vascular dysfunction," Press release. (April 3, 2005).

American Urological Association. "*Urinary Tract Infections in Adults*." www.Urology-Health.org. Accessed 2/26/2006.

Angier, Bradford. "Field Guide to Medicinal Wild Plants." Mechanicsburg, PA: Stackpole Books, 1978.

Barnes, N. B., et al. "Consumers, Cranberries, and Cures: What Consumers Know About The Health Benefits of Cranberries," Slade's Ferry Bank Center for Business Research (Spring 2002).

Burger, Ora, et al. "A high molecular mass constituent of cranberry juice inhibits *Helicobacter pylori* adhesion to human gastric mucus." *FEMs Immunology and Medical Microbiology* 29 (December 2000) 295.

Castleman, Michael. *Nature's Cures*. Emmaus, PA: Rodale Press, 1996.

Chu, Y. F., Liu, R. H. "Cranberries inhibit LDL oxidation and induce LDL receptor expression in hepatocytes." *Life Sciences* 77, no. 15 (August, 26, 2005) 1892–901.

Cranberry Institute, The. "Cranberry Nutritional Composition." www.cranberryinstitute.org.

Crews, W. David Jr., et al. "A Double-blind, Placebo-controlled, Randomized Trial of the Efficacy of Cranberry Juice in a Sample of Cognitively intact Older Adults: Neuropsychological Findings. *Journal of Alternative and Complementary Medicine* 11, no. 2 (April 2005) 305–09.

Duyff, Roberta Larson. *The American Dietetic Association's Complete Food and Nutrition Guide*. Minneapolis: Chronimed Publishing, 1996, 1998.

Erbringer, A., Rashid, T. "Rheumatoid Arthritis Is an Autoimmune Disease Triggered by Proteus Urinary Tract Infection." *Clinical and Developmental Immunology* 13, no.1. (March 2006) 41–48.

Ferguson, P. J. et al., "A Flavonoid Fraction from Cranberry Extract inhibits Proliferation of Human Tumor Cell Lines." *Journal of Nutrition* 134 (June 2004) 1529–535.

Foster, Steven, Tyler, Varro E. *Tyler's Honest Herbal: A Sensible Guide to the Use of Herbs and Related Remedies.* Binghampton, NY: Hawthorne Herbal Press, 1999.

Gettmann, Matthew T., et al. "Effect of Cranberry Juice Consumption on Urinary Stone Risk Factors." *The Journal of Urology* 174, no. 2 (August 2005) 590–94.

Head, E., (report) "Ninth International Conference on Alzheimer's Disease and Related Disorders, July 17–22, 2004, Philadelphia, PA." *Brain Aging Bulletin* 5, no. 1 (Winter 2004).

Heinerman, John, PhD, *Folk Remedies from Around the World: Traditional Cures for 300 Common Ailments*. New York: Prentice Hall, 2000.

Howell, Amy B., et al. "A-type Cranberry Proanthocyanidins and Uropathogenic Bacterial Anti-adhesion Activity." *Phytochemistry* 66, no.18 (September 12, 2005) 2281–91.

Infectious Diseases Society of America. Press Release: "IDSA Releases Hit List of Dangerous Bugs." (March 1, 2006).

Jepson, R. G., et. al., "Cranberries for Preventing Urinary Tract Infections." *The Journal of Urology* 173, no. 1 (January 2005) 111–12.

Johnson-White B., et al. "Prevention of Nonspecific Bacterial Cell Adhesion in Immunoassays by Use of Cranberry Juice." *Annals of Chemistry* 78, no. 3 (3) (February 1, 2006) 823–27.

Jones, A. H. "The Next Step in Infectious Disease: Taming Bacteria." *Medical Hypotheses* 60, no. 2 (February 2003) 171–74.

Joseph, James A., Nadeau, Daniel A., Underwood Anne. *The Color Code: a Revolutionary Eating Plan for Optimum Health.* New York: Hyperion, 2002.

Kim, T. S., et al. "Decreased Plasma Antioxidants in Patients with Alzheimer's Disease." *International Journal of Geriatric Psychiatry* 21, no. 4 (April 2006) 344–48.

Koh, S. H., et al. "The Effect of Epigallocatechin Gallate on Suppressing Disease Progression of ALS Model Mice. *Neuroscience Letters* 395, no. 2 (2) (March 6, 2006) 103–07.

Koo, H., et al. "Influence of Cranberry Juice on Glucan-Mediated Processes Involved in *Streptococcus mutans* Biofilm Development." *Caries Research*, vol. 40. no. 1, 2006.

Kurutas, Ergul Belge, et al. "The Effects of Oxidative Stress in Urinary Tract Infection." *Mediators of Inflammation* 4 (2005) 242–44.

Lau, F. C., et al. "The Beneficial Effects of Fruit Polyphenols on Brain Aging. *Neurobiology of Aging* Suppl. 1 (December 2005) 128–32.

Lin, Y. T., et al. "Inhibition of *Helicobacter pylori* and Associated Urease by Oregano and Cranberry Phytochemical Synergies." *Applied and Environmental Microbiology* 71, no. 12 (December 2005) 8558–564.

Lynch, D. M., "Cranberry for Prevention of Urinary tract Infections," *American Family Physician* 70, no. 11 (December 1, 2004).

Marambaud, P., et al. "Resveratrol Promotes Clearance of Alzheimer's Disease Amyloid-beta Peptides." *Journal of Biological Chemistry* 280, no. 45 (November 11, 2005) 37377–382.

McBride, Judy. "High ORAC Foods May Slow Aging." USDA Agricultural Research Service (February 8, 1999) www.ars.usda.gov/is/pr/1999/990208.htm Accessed on Feb. 27, 2006.

Montiel, T., et al. "Role of Oxidative Stress on Beta-amyloid Neurotoxicity Elicited During Impairment of Energy Metabolism in the Hippocampus: Protection by Antioxidants." *Experimental Neurology.* Epub ahead of print (April 18, 2006).

National Kidney and Urologic Disease Information Clearinghouse (NKUDIC) "Urinary Tract Infections in Adults." http://kidney.niddk.nih.gov.

Ono, K., et al. "Anti-amyloidogenic Effects of Antioxidants: Implications for the Prevention and Therapeutics of Alzheimer's disease." *Biochimica et Biophysica ACTA.* Epub ahead of print (April 7, 2006).

Puupponin-Pimia, Riita, et al. "Bioactive Berry Compounds—Novel Tools Against Human Pathogens." *Applied Microbiology and Biotechnology* 67, no. 1 (April 2005).

Reed, J. "Cranberry Flavonoids, Atherosclerosis, and Cardiovascular Health." *Critical Reviews in Food Science Nutrition* 42, no. 3, supplement (2002) 301–16.

Rimando, A. M., et al. "Resveratrol, Pterostilbene, and Piceatannol in *Vaccinium* Berries." *Journal of Agriculture and Food Chemistry* 52, no. 15 (July 28, 2004) 4713–719.

Scalbert, Augustin, et al. "Dietary Polyphenols and Health: Proceedings of the 1st International Conference on Polyphenols and Health." *The American Journal of Clinical Nutrition* 81, no. 1 (January 2005) 215S–217S.

Schardt, David. "Berry Berry Good?" *Nutrition Action Health Letter*, 32, no. 5. (June 2005).

Schieszer, John. "Cranberry Products May Prevent UTIs: Ingestion of Cranberry Capsules Suppresses Three Common Pathogens in Urine Specimens." *Renal Urology News* 4, no. 10 (November–December 2005) 11.

Seeram, N. P., et al. "Total Cranberry Extract versus Its Phytochemical Constituents: Antiproliferative and Synergistic Effects against Human Tumor Cell Lines," *Journal of Agriculture and Food Chemistry* 52 (2004) 2512–517.

Shmuely, H., et al. "Susceptibility of *Helicobacter pylori* Isolates to the Antiadhesion Activity of a High-molecular-weight Constituent of Cranberry." *Diagnostic Microbiology and Infectious Diseases* 50, no. 4 (December 2004).

Sifton, D. W., editor in chief, "Putting an End to Urinary Tract Infections" in *The PDR Family Guide to Women's Health and Prescription Drugs.* Montvale, NJ: Medical Economics Co., January 2003. Accessed February 15, 2006 at: http://wf2la2.web feat.org/4gS9F12/url+http://galenet.galegroup.com.

Sims, J. B., et al. "Urinary Tract Infection in Patients with Acute Coronary Syndrome: A Potential Systemic Inflammatory Connection." *American Heart Journal* 149, no. 6 (June 2005) 1062–065.

Subak, Geneil J., Herbert, Victor., eds. *The Mount Sinai School of Medicine Complete Book of Nutrition.* New York: St. Martin's Press, 1990.

Sun, J. et al. "Antioxidant and Antiproliferative Activities of Common Fruits." *Journal of Agriculture of Food Chemistry.* 50, no. 25 (December 4, 2002) 7449–454.

Tapiero, H., et al. "Polyphenols: Do They Play a role in the Prevention of Human Pathologies?" *Biomedicine and Pharmacotherapy* 56, no.4 (June 2002) 200–07.

University of Georgia: Cranberry—*Vaccinium macrocarpon* www.uga.edu/fruit/cran beri.htm. Accessed on Jan. 27, 2006.

University of Massachusetts Cranberry Station: "Natural History of the American Cranberry." www.umass.edu/cranberry/cranberry/seasons.shtml. Accessed 02/07/2006.

University of Massachusetts Dartmouth. "New Study Finds Cranberry Compounds Block Cancer: First Study to Confirm Cranberry Proanthocyanidins Inhibit Growth of Tumor Cells." Press Release. *PR Newswire.* January 25, 2006.

University of Virginia Health System. "Damage From Oxygen May Be One Cause of Parkinson's Disease." Press Release. *Newswise.* May 16, 2006.

University of Wisconsin–Madison. "The American Cranberry (*Vaccinium macrocar-pon*), Frequently Asked Questions. www.library.wisc.edu/guides/agnic/cran berry/faq.htm Accessed Feb. 7, 2006.

Vattem, D. A., et al. "Synergism of Cranberry Phenolics with Ellagic Acid and Ros-marinic Acid for Antimutagenic and DNA Protection Functions." *Journal of Food Biochemistry* 30, no. 98 (February 2006).

Weil, Andrew. *Eating Well for Optimum Health: The Essential Guide to Food, Diet, and Nutrition.* New York: Knopf, 2000.

Weiss, E. I., et al. "Inhibitory Effect of a High-molecular-weight Constituent of Cran-berry on Adhesion of Oral Bacteria." *Critical Reviews in Food Science and Nutri-tion* 42, no. 3, suppl. (2002) 285–92.

Weiss, E. L., et al. "Inhibiting Interspecies Coaggregation of Plaque Bacteria with a Cranberry Juice Constituent." *Journal of the American Dental Association* 130, no. 1 (January 1999) and 130, no. 3 (March 1999).

Weiss, E. I., et al. "Cranberry Juice Constituents Affect Influenza Virus Adhesion and Infectivity." *Antiviral Research* 66, no. 1 (April 2005) 9–12.

Zhanel, G. G., et al., "Antibiotic Resistance in Outpatient Urinary Isolates: Final Results from the North American Urinary Tract Infection Collaborative Alliance (NAUTICA)," *International Journal of Antimicrobial Agents* 26, no. 5 (December 1, 2005) 380–88.

Zhang, L. et al., "Efficacy of Cranberry Juice on *Helicobacter pylori* infection: a Dou-ble-Blind, Randomized Placebo-Controlled Trial." *Helicobacter* 10, no. 2, (April 2005) 139–145.